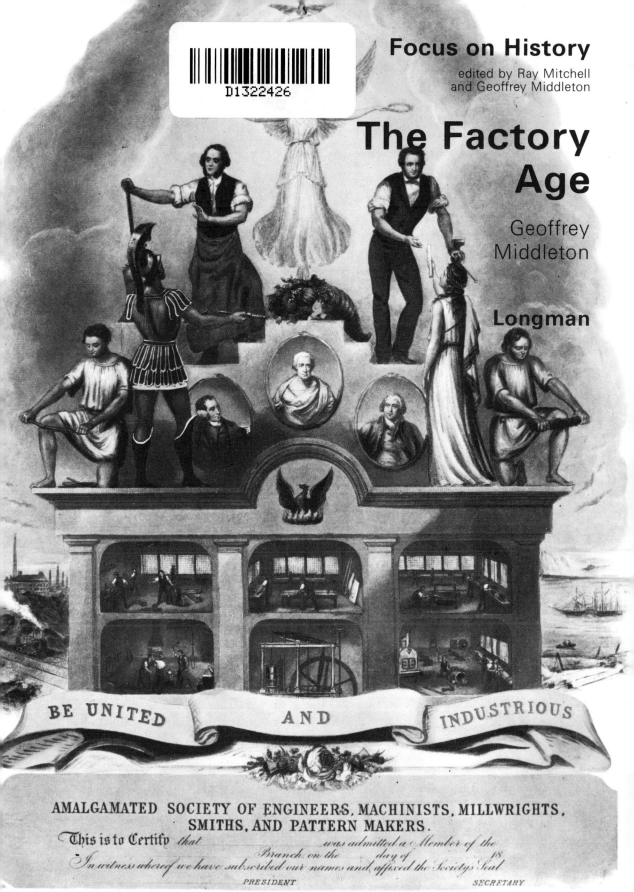

Focus on History
edited by Ray Mitchell
and Geoffrey Middleton

The Factory Age

Geoffrey Middleton

Longman

BE UNITED · AND · INDUSTRIOUS

AMALGAMATED SOCIETY OF ENGINEERS, MACHINISTS, MILLWRIGHTS, SMITHS, AND PATTERN MAKERS.

This is to Certify that ___ was admitted a Member of the ___ Branch, on the ___ day of ___ 18 ___ In witness whereof we have subscribed our names and affixed the Society's Seal

PRESIDENT   SECRETARY

This is a picture of the industrial part of Leeds during the 19th century. Can you see:
— the large factories, or mills, as they were called. Here hundreds of men, women and children worked for most of their waking hours.
— the countless factory chimneys stretching away into the distance. Notice the thick, black smoke coming from many of them.
— the jumble of houses in the centre of the picture and the street of terraced houses on the left. Some have not even a backyard of their own.
— the lines of washing strung across a patch of bare ground. What will stop the washing from keeping clean as it dries?

This book is about a period of time when new inventions changed people's lives more rapidly than ever before. It became known as the Industrial Revolution, or the Factory Age. Leeds was only one of many towns where large factories sprang up and poor, small houses were quickly built for the many workers needed to use the new machines.

The shaded time-chart below shows only approximately when this happened, for it did not begin or end at an exact date. Just as our own way of life nowadays is constantly changing, so men have always made new discoveries and inventions. Your parents and grandparents will be able to tell you of many which have been made since they were children.

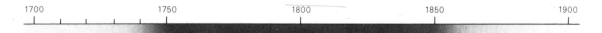

| 1700 | 1750 | 1800 | 1850 | 1900 |

So our story begins about the middle of the 18th century. At that time we believe there were about eight million people living in Great Britain. This is only an approximate figure for there was no census, or official count of people, until 1801.

Below is a picture of a village about that time. Most people still lived in villages near to a market town and many worked on the land as their fathers and grandfathers had done before them. Some landowners had enclosed their land with hedges or fences and had begun to use new methods of farming. You can read about these in another *Focus on History* book called *Georgian England.*

For many years most industries had been based in the workers' own homes and they were called 'cottage industries'.

The cottage industries

Since the Middle Ages the manufacture of woollen cloth had been Britain's main industry and export. Although it was woven almost everywhere in the country, this map shows where most of it was made.

There were few machines and factories as we know them today. Wealthy merchants bought the raw wool from the farmers and brought it to the villagers to be spun into yarn and then woven into cloth. The spinning and weaving was carried out in the villagers' cottages.

In the picture below a woman is at work in her cottage. She is spinning or twisting the 'rovings' of wool into yarn on her spinning wheel. Find the stiff wire brushes on the floor. These were called 'carders' and the little girl used them to brush the raw wool into separate fibres. She probably learned to do this when very young. Notice how the spinning wheel has been placed near the cottage window so that the woman can work in good light.

Look again at the picture and find:
— the spindles of spun yarn in the baskets on the floor
— the fireplace, which was used for cooking as well as heating
— the heavy, iron kettle.

Notice the clothes the woman and the little girl are wearing.

Weaving was strenuous and more highly skilled work than spinning, and was nearly always carried out by men. The picture above shows the long threads being woven into cloth on a loom. The weaver worked the loom with his foot treadles.

One writer of this period described a typical weaver like this:

> . . . his green woollen apron twisted round his waist, his clean shirt showing at the open breast of his waistcoat, his brown silk handkerchief wrapped round his neck, a quid (piece) of tobacco in his mouth, and a broad and rather slouched hat on his head.'

Would this be an accurate description of the weaver in the picture above? If not, write your own description of him.

The woven cloth was collected by the merchant, who paid the spinners and weavers for their work. Sometimes they were members of the same family and sometimes families joined together to do the work. Some workers were full-time craftsmen, but others worked in the evenings after a hard day's work in the fields.

Next, the cloth was prepared for the market by fullers, bleachers and dyers, and sold for use in other parts of the country, or exported overseas.

Here is a picture of two dyers at work. It is taken from a woodcut of the early 19th century, but dyers were working in the same way many years earlier.

Look for:
— the man soaking the cloth in the tub of dye
— the hand water-pump
— the pile of cloth awaiting dyeing
— the man hanging out the dyed pieces of cloth to dry.

At first, vegetable dyes were mostly used, but new colouring matters were introduced during the 18th century.

The people in this picture lived in Wensleydale in Yorkshire and had their own industry. Can you see what each one of them is doing?

It is said that they knitted, not only in their spare time, but whenever they had no need to use their hands for their normal work. So the man is knitting as he drives the sheep along the road. A man could knit as he kept watch over his flock, and one woman is said to have knitted a pair of men's stockings each day on her way to and from market. There she would sell those knitted garments the family knitters did not need for their own use.

Look how the people are dressed and notice particularly:
— the buckled shoes, breeches and stockings of the men
— the bulky layers of long clothes of the women
— the soft hats worn by everyone except the little girl.

In some places country people organised their own industries. These existed side by side with the other cottage industries which were dependent upon the merchants. But changes were on the way.

7

Rises in population

The population was beginning to increase rapidly. Look at this chart which shows the rise of population in Great Britain between 1750 and 1851, a period of just over a hundred years. Remember that the figures are only approximate until 1801. Can you recall why? If not, read the first paragraph of page 3 again.

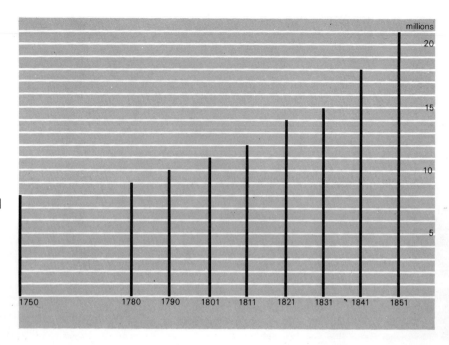

millions

20

15

10

5

1750 1780 1790 1801 1811 1821 1831 1841 1851

How many people lived in Great Britain:
— in 1750?
— in 1780? (30 years later)
— in 1790? (40 years later)
— in 1801? (51 years later).

By how much had the population increased since 1750?

By which year was the population double that of 1750? How many years had it taken to double?

By how much had the population increased between 1750 and 1851?

During which ten years was there the largest increase in the population?

Now begin to make your scrapbook about the Factory Age. In it you can write notes about the information you discover and include some coloured maps and illustrations. Here are some ideas for the first part of your book.

Find out what life was like in the middle of the 18th century in the district in which you live. Try to discover the population of your village or town, the names of well-known local people at that time and any changes which were taking place. Look for a map of your district and copy it. Your local reference library or museum may be able to help you to find this information.

The beginning of the Factory Age

As the population increased more goods were needed, but the workers were unable to make enough in their own homes or small workshops. New ways had to be found to produce the goods cheaply and more quickly.

New machines were invented, first of wood and then of iron. They were placed, not in the workers' homes, but in factories, where more goods could be produced more quickly. So work moved from houses to factories.

New ways of working machines were invented, at first by water instead of by hand, and later by steam when steam power was discovered. For steam more coal was needed and this led to better methods of producing coal and iron. All this again led to further inventions — railways and steamships which carried goods more quickly from place to place. So we moved into the Factory Age.

This simple chart shows how some of the many changes which occurred during this Factory Age depended on each other. Copy it into your scrapbook and also make a time-chart from 1750–1850, on which you can record, by means of illustrations and notes, the various developments as they came during those years.

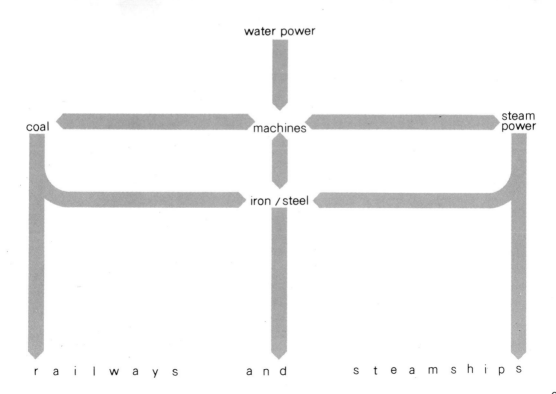

Textile inventions

1733   A flying shuttle was invented by John Kay. It enabled a weaver to weave wider cloth on his loom than before, without the aid of an assistant. It speeded up the weaving process and increased the demand for yarn. So better spinning machines were required.

1767   On the left is the 'spinning jenny' invented by James Hargreaves. It could spin eight threads at once. Later models increased the number of threads to 120. The cottage spinner could rapidly increase his supply of yarn.

1769   The spinning frame on the right was invented by Richard Arkwright. Only four threads were spun at a time, but these were stronger than those made on the spinning jenny. This machine was powered by a horse which walked round in a circle, turning a wheel. Later models were installed in riverside factories where the rushing water turned a water-wheel which drove the machinery. So Arkwright's machine became known as a 'water frame'.

Arkwright's machine could spin only hard and strong yarns, and softer yarns were still produced on the spinning wheel or 'jenny'. Both kinds of yarn were needed and were woven together by the weaver on his loom. But only the cheaper kind of materials could be woven from the coarse, machine-spun yarns.

1779   The machine above called a 'mule', was invented by Samuel Crompton, a Bolton spinner. It spun thread much finer than machines had ever done before. So it was now possible to spin, by machine, fine as well as coarse yarns.

The new spinning machines produced better quality yarns much faster, at a lower cost than ever before, and were in great demand. But it soon became impossible for the hand looms to cope with all the yarn produced.

1785　A power loom was
invented by
Edmund Cartwright,
but it was not
really successful
until various
improvements were
made by other
inventors.

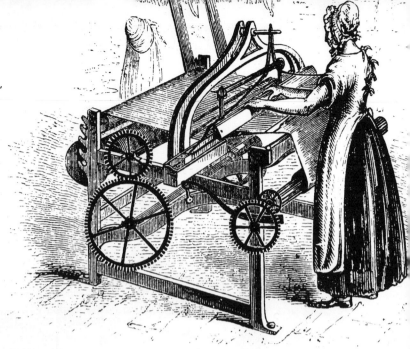

In this picture a
woman is using a
power loom about
30 years after
Cartwright's
invention. Compare it with the picture on page 5.

Women were able to look after the new power looms, for all the operations were carried out mechanically. Their main task was to watch for threads breaking, and, when they did, to stop the machine to repair them.

This chart shows how rapidly the improved power looms were brought into use. Can you work out approximately how many power looms were in use about 30 years, 50 years and 60 years after Cartwright's invention?

But the new machines did not help everyone, for although some of the patterned weaving was still done on hand looms in the cottages, there was less work available for the skilled hand loom weavers and many became out of work. Those employed could earn only a few shillings a week, for they were glad to accept work even at low wages.

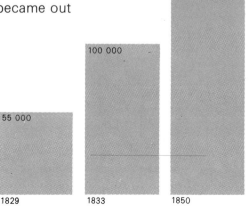

2400	14 150	55 000	100 000	250 000
1813	1820	1829	1833	1850

This picture shows some very skilled Yorkshire workers called croppers. They are cutting the nap, or rough surfaces, off the woollen cloth with large shears to make it smooth. The weight of the shears could be increased by fixing lead weights to them, and the greater the weight of the shears the smoother the cloth became. You can see two weights on the table.

Here is a shearing machine which could do the work of four croppers. The roll of woollen cloth passed over a cylinder, on which was fixed a sharp blade. As the cylinder turned, the blade cut the nap off the cloth.

When these shearing machines were installed in the factories, many croppers had no work.

The Luddites

In the early 19th century there was much unemployment, wages were low, and heavy taxation to pay for the wars against France had raised the cost of living. Some workers formed gangs to protest against the hardships they were suffering. Many complained about the poor pay, while others said the new machines were being used unfairly by their employers.

Some of the gangs broke into factories at night and smashed certain machines. In Nottingham they destroyed the frames on which wide cloth was being knitted and cut into strips for stockings. This was cheaper for the employers than using the stocking frames. There were similar riots in other areas, so sometimes soldiers were brought in to protect the factories and Parliament passed an Act which made machine-breaking punishable by death.

These gangs became known as Luddites, for it was thought they were led by someone called Ned Ludd. But it is not certain that Ned Ludd really existed. Here is part of a letter supposed to have been written by Ned Ludd and sent to Mr. Smith, the owner of some shearing frames in Yorkshire.

> Sir
> - Information has just been given in that you are a holder of those detestable Shearing Frames; and I was desired by my Men to write to you and give you fair Warning to pull them down, and for that purpose I desire you will now understand I am now writing to you you will take Notice that if they are not taken down by the end of next Week, I will detach one of my Lieutenants with at least 300 Men to destroy them and furthermore take Notice that if you give us the Trouble of coming so far we will increase your misfortune by burning your Buildings down to Ashes and if you have Impudence to fire upon any of my Men, they have orders to murder you, burn all your Housing, you will have the Goodness to your Neighbours to inform them that the same fate awaits them if their Frames are not speedily taken down as I understand their are several in your Neighbourhood. Frame holders. And as the Views and Intentions of me and my Men have be...

What does Ned Ludd tell Mr. Smith to do?
What does he threaten to do to:

- the shearing frames and the building
- Mr. Smith and the neighbours?

How do workers protest against their working conditions nowadays?

Water power

Although some hand spinning and weaving continued in the cottages, it was soon discovered that water power could be used to drive the new machines. The rushing streams of water tumbling down the slopes of the Pennines were strong enough to turn large water-wheels and drive the machines, and there was plenty of water available for dyeing and bleaching. The damp atmosphere and fairly even temperature of Lancashire was ideal for cotton spinning and the raw cotton from America came in through the nearby port of Liverpool. From Liverpool, too, goods could be exported overseas.

So merchants who had money to invest, built large mills, or factories, to house the new machines along the banks of the rivers in Lancashire, Yorkshire and Derbyshire, and workers moved into the towns to look for work in the new factories.

TO
Journeymen Spinners

Wanted Immediately,
From Eighty to One Hundred
MULE SPINNERS,

For a New Mill and other Mills, in Great Bolton, which New Mill is complete with new Machinery now ready gaited, and will commence running on Monday Morning next, adjoining to which Mills are a Number of Cottages, for the convenience and accommodation of Spinners: liberal Wages will be given and constant employ.

For further particulars apply to Messrs. ORMROD and HARDCASTLE, of Bolton aforesaid, Cotton Spinners.

BOLTON, 7th NOVEMBER, 1816. [EDWARD SMITH, PRINTER.]

Look at the advertisement above for spinners in a new mill at Bolton in Lancashire. What three things did these mill owners offer to persuade workers to apply for jobs? Try to find out what 'journeymen' were.

The mill owners employed managers and foremen to supervise the many workers in the new mills. The factory towns rapidly grew larger, industries became better organised, methods of transport improved and new markets opened up at home and abroad.

This huge water-wheel at a cotton mill near Bolton cost £5,000, which was an enormous sum of money at that time. Compare the size of the wheel with the height of the people standing beside it. The wheel was said to be equal to 150 horse power.

But water power was not always satisfactory. Factories could be built only where there was a supply of rushing water to turn the water-wheels. When the rivers were low the wheels turned only slowly and sometimes in hot weather the rivers ran dry and all work ceased.

One Lancashire weaver wrote in his diary:

May 29 'Another very warm day, and this dry weather is much against us as the Ribble is very low and in the afternoons our looms go very slow for want of water.'

June 7 'We were stopped nearly all day for want of water.'

August 28 'There were 30 mills stopped in Blackburn this week for want of water, and will not start again until wet weather sets in.'

The great need of mills was power, so where water power was unsatisfactory other sources had to be found.

The coming of steam power

During the 17th century, several men had tried to discover how the power of steam could be used. One of them, Thomas Savery, invented a steam pumping-engine which was described as:

'A new invention for Raising Water and Occasioning Motion to all Sorts of Mill Work by the Impellent Force of Fire, which will be of great use and Advantage for draining Mines, serveing Towns with Water, and for the working of all Sorts of Mills where they have not the Benefit of Water nor Constant Windes.'

But Savery's engine had only limited uses and was liable to explode unless expertly used.

The Engine to raise Water by Fire

In 1712, after many experiments, Thomas Newcomen built this steam engine for pumping water out of mines.

Look closely at the picture and find:
- the large wooden beam which acted as the pump handle
- the chains at the right of the beam fastened to pump rods which went deep down the mine
- the chains at the other end of the beam which fastened to a piston
- the long cylinder into which the piston fitted
- the brick boiler and the fire underneath
- the water tank on the wooden beams above the cylinder.

Now look at the diagram below of Newcomen's engine and try to follow how it worked.

The weight of the pump rods going down the mine pulled the piston to the top of the cylinder. Water in the boiler was heated by a fire underneath and turned into steam, which entered the cylinder through a valve so that it could not return to the boiler. (Rather like the way in which air enters your cycle tyres.)

When steam filled the cylinder it was condensed (changed back into water) into a smaller space by a jet of cold water from the tank. This caused a vacuum, and pressure of air from above pushed the piston to the bottom of the cylinder, and buckets at the end of the pump rods lifted the water out of the mine.

A number of Newcomen's engines were used throughout the 18th century. They pumped water from coal and tin mines, raised water into mill reservoirs to turn the water-wheels, and pumped water supplies into some of the towns. But they were expensive to build, worked slowly, and were noisy. They used a tremendous amount of fuel — some as much as 13 tons of coal a day — and often they went wrong.

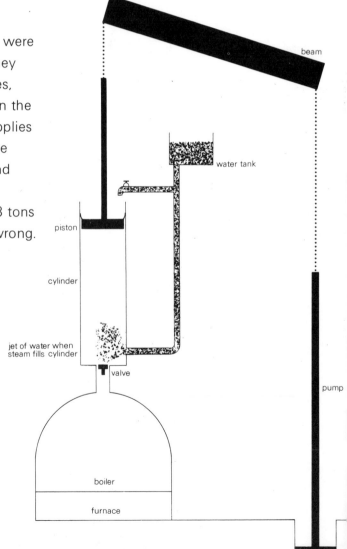

beam

water tank

piston

cylinder

jet of water when steam fills cylinder

valve

pump

boiler

furnace

A Scotsman, James Watt, was one of several engineers who tried to improve Newcomen's engines. Here is his portrait. It can be seen in the National Portrait Gallery in London.

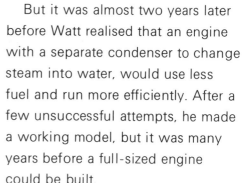

James Watt was born at Greenock in 1736 and was trained as an instrument maker. In 1763 he was asked to repair the model of a Newcomen engine shown below. This model is now in the Hunterian Museum, Glasgow University. While he was repairing it, Watt discovered that steam heat was being wasted with each stroke of the engine.

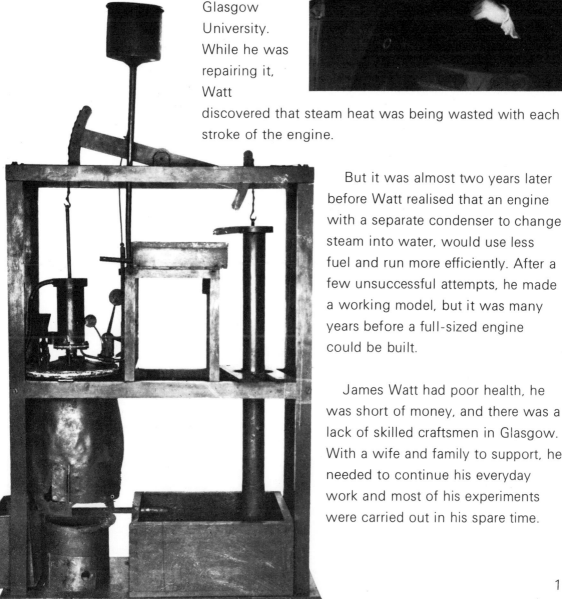

But it was almost two years later before Watt realised that an engine with a separate condenser to change steam into water, would use less fuel and run more efficiently. After a few unsuccessful attempts, he made a working model, but it was many years before a full-sized engine could be built.

James Watt had poor health, he was short of money, and there was a lack of skilled craftsmen in Glasgow. With a wife and family to support, he needed to continue his everyday work and most of his experiments were carried out in his spare time.

Here is a large factory at Soho, near Birmingham, where metal toys, trinkets and ornaments were being made by skilled craftsmen. It was owned by a businessman called Matthew Boulton. Notice the well-dressed visitors to the factory. Some are travelling in fine carriages and one is on horseback. Perhaps they are wealthy customers coming to discuss business with Matthew Boulton.

In 1774 James Watt needed financial help and left Glasgow to become a partner of Matthew Boulton. Within two years an engine with a separate condenser was built. This was still only a pumping engine, but in 1781 Watt built an engine which changed the up and down pumping action of earlier steam engines to one which could turn, or rotate, a large fly-wheel.

Watt made further improvements to these rotative engines so that belts attached to the fly-wheels could drive the machinery in textile mills, potteries, breweries, flour mills and many other industries. This was very important, for factories were no longer dependent upon water power. They began to be built close to the coalfields as coal was needed to drive the steam engines. Watt's engines were in such demand that he became a very wealthy man.

Some of his engines are now in the Science Museum in London and the one on the right can be seen working at various times each day. Look at the picture and notice the toothed gear-wheels and the large fly-wheel.

This is James Watt's attic workshop, which was his home at Heathfield near Birmingham from 1790 until his death in 1819. In 1824 it was dismantled and reassembled at the Science Museum in London where it can be seen today.

Find:
— the stove on the left used for heating the room and for some of Watt's experiments
— Watt's workbench in front of the window
— his lathe which was worked by a foot treadle in the centre of the picture.

Now look at the shelves on the right of the workshop. Find:
— the jars of chemicals Watt used for his experiments
— his leather apron hanging up
— the small statuette on the top shelf.

James Watt's rotative engine showed men how to use the power of steam to drive all kinds of machinery. By 1800, about 500 engines had been set up, some in cotton mills, some in coal mines, some on canals and some in breweries. Britain began to change from an agricultural country into an industrial one. Large supplies of coal were needed for the steam engines. New factories, as well as iron foundries, were built near the coalfields. More people moved from the country into the towns to find work.

Iron and steel

At the beginning of the 18th century iron could be produced only by melting iron ore in a charcoal furnace. So the ironworks were close to wooded countryside where there was a plentiful supply of timber for charcoal. They were built on the banks of fast moving rivers so that water power could pump blasts of air through the furnaces. There were ironworks in Sussex, Shropshire and South Wales.

The picture above shows an ironworks in the 18th century at Coalbrookdale in Shropshire, on the banks of the River Severn. It was used for casting and boring cannon. Notice:
— the furnaces and the tall smoking chimneys
— the cannon lying on the river bank near the boring mill
— the man pointing out the large water-wheel to the boy who is fishing. It turned the drill which bored a hole through the cannon.

Early in the 18th century Abraham Darby discovered how to use coke, made from coal, to smelt the iron at his Coalbrookdale works. The discovery came at a time when there was a growing shortage of timber for charcoal burning and led to an increase in iron-making. In 1749 Darby's son discovered how to make better quality bar-iron and this, too, increased the supply of iron wherever coal and iron ore could be found close together.

Here is an enlarged photograph of a token coin — value one halfpenny — issued by the Coalbrookdale Company. What is the date of the coin? Coins like these were issued for wages because of the shortage of small change towards the end of the 18th century and could be spent locally.

Look at the bridge shown on the coin. When was it built? It was the first iron bridge in the world and was cast in the Coalbrookdale Ironworks at the time of Abraham Darby's grandson, who was also called Abraham. For many years, people came from far and near to see this famous bridge, and those who crossed it by carriage had to pay a toll of one shilling. In the Shropshire Record Office we can read this description of the bridge written in 1801:

> . . . you arrive at the banks of the River Severn, where you have a full view of the Iron Bridge. The noble Arch presenting itself to the inspection of the curious traveller; whose massy curved ribs fill him with astonishment; . . . the ribs, cover'd with iron plates and connected with strong pillars, will prove an object at once majestic and beautiful.'

The town of Ironbridge grew up around the bridge and took its name from it. This is how the bridge looks today, almost two hundred years since it was built.

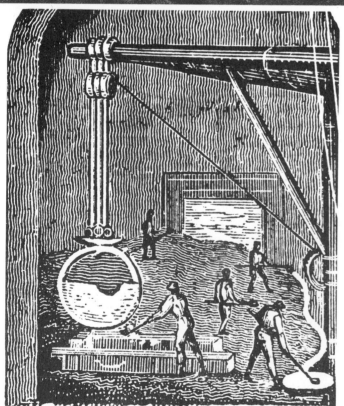

The picture above shows iron ore being melted to separate the metal from the rock. This was called smelting. Notice the molten iron flowing out of the glowing white-hot iron blast furnace and being guided into channels. Sometimes the molten iron was run directly into moulds of the shapes required, but often the 'pigs' or rough shapes of iron were re-melted in a foundry and re-cast into moulds, as shown in the picture on the left.

Find the man who is skimming the molten iron to remove the scum as it flows from the furnace into a basin. Another man with a hand ladle is stirring the metal before transferring it to the great ladle on the left, which is worked by an overhead crane. Under the great ladle are the moulds in which the castings are made.

In 1776 John Wilkinson used a steam engine to blow his blast furnace, instead of a water-wheel, and in 1784, Henry Cort and Peter Onions both discovered a method, known as 'puddling', for producing high quality iron quickly and cheaply.

Many iron foundries sprang up close to the coalfields, for they needed vast quantities of coal. Newcastle, Birmingham and Sheffield became the leading iron-manufacturing areas in the country and this diagram shows how the yearly production of iron rapidly increased.

17 000 Tons	68 000 Tons	250 000 Tons	1 000 000 Tons	2 000 000 Tons	
1740	1790	1800	1835	1850	

Huge quantities of iron were needed to make all kinds of factory machinery, to build bridges, and, a little later, for locomotives, rails and steamships.

Some iron was refined to make steel, especially around Sheffield, where it was cut and ground into various tools and also made into cutlery. Below is a picture of a workshop at a steelworks in Sheffield where workmen are making files. What is the

man on the left doing? The three work-men near the window are hammering the ridged pattern on the files. Can you find a pile of finished files in the picture?

t wo
PENCE
a pair

G. R.

WE whose Names are here *under-written, Justices of the Peace for the County of Essex aforesaid (whereof one is of the Quorum) Do consent to the putting forth the said Elizabeth Smith — an Apprentice according to the Intent and Meaning of this Indenture.*

Wm Smith

T G Hramston

His Indenture

in the *fortieth* —— Year of the R

by the Grace of God of *Great Britain*, Fr

in the Year of our Lord 1799 **Witne**

David Wood —— Church-

of *Essex* **And**

Matthew Newcome —

parish by and with the Consent of h

whose Names are hereunto subscribed, ha

Elizabeth Smith —— age

Apprentice to *John Douglas of Per*

of Lancaster Esquire Cotton Worker and Manuf

dwell and serve from the Day of the Date of these Presents, until the said A

Marriage which shall first happen according to the Statute in that Case made and p

said *Master* faithfully-shall serve in all lawful Businesses, according to

in all Things demean and behave her self towards her said *Master*

Douglas for himself, his Executors and Ad

Wardens and Overseers, and every of them, their, and every of the

Successors, for the Time being, by these Presents, That he the said

the said Apprentice in *the Craft Mistery*

shall and will teach, and instruct, or cause to be taught and instructed in th

thereunto belongeth or in any wise app

the Term aforesaid, find, provide, and allow unto the said Apprentice

Washing, and other Things necessary and fit for an Apprentice: (**Provid**

John Douglas —— his Executors and Administrators, to

longer Time than Three Calendar Months next after the Death of the said

Douglas shall happen to die during the Continuance of such App

the Thirty-second Year of the Reign of King George the Third, inti

tices,") **And** also shall and will so provide for the said Apprentice, that

Parishioners of the same; but of and from all Charge shall and will save th

nified during the said Term.

Signed Sealed and Delivered in the
Presence of

In Witness whereof, the Parties abovesaid to these p

the Day and Year first above written.

Thos Anderson

Wm Riggs

26

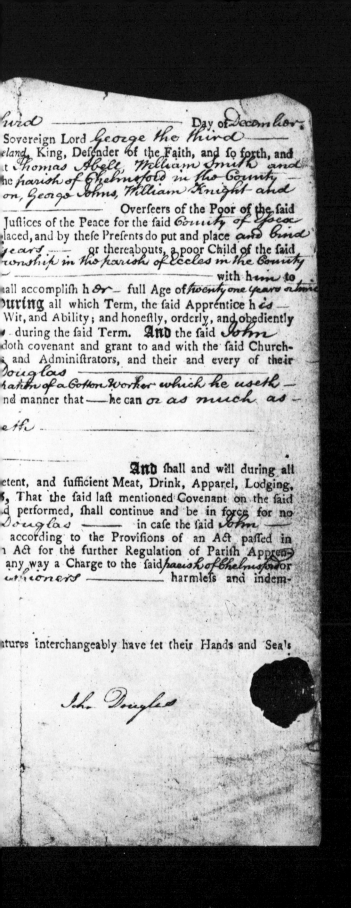

hird ——————————— Day of December
Sovereign Lord *George the third* ————
eland, King, Defender of the Faith, and so forth, and
t *Thomas Abell, William Smith* and
he *parish of Chelmsford in the County* ——
on, *George Johns, William Knight* and
———————— Overseers of the Poor of the said
Justices of the Peace for the said *County of Essex*
laced, and by these Presents do put and place *and bind*
years ———— or thereabouts, a poor Child of the said
Township in the parish of Eccles in the County
—————————————————— with *him* to
all accomplish h *or* — full Age of *twenty one years a time*
During all which Term, the said Apprentice h *is* —
Wit, and Ability; and honestly, orderly, and obediently
— during the said Term. **And** the said *John*
doth covenant and grant to and with the said Church-
and Administrators, and their and every of their ——
Douglas
hation of a Cotton Worker which he useth ——
nd manner that —— he can *or as much as* ——

eth ————

———————— **And** shall and will during all
etent, and sufficient Meat, Drink, Apparel, Lodging,
, That the said last mentioned Covenant on the said
d performed, shall continue and be in force for no
Douglas ———— in case the said *John* ——
according to the Provisions of an Act passed in
Act for the further Regulation of Parish Appren-
any way a Charge to the said *parish of Chelmsford* or
missioners ———————— harmless and indem-

atures interchangeably have set their Hands and Seals

John Douglas

Inside the factories

Much of the work in the factories was done by women and children. Sometimes the factory owners arranged for orphans and poor children to be sent from workhouses in the south to work in the northern mills as apprentices. This relieved the parish of the expense of these children and provided the employers with cheap labour.

The document on the left is an agreement called an Indenture. It bound one of these children, called Elizabeth Smith, to work in a Lancashire cotton mill. The original document is held in the Essex County Record Office, but you can examine this copy with the aid of a magnifying glass. Can you find the answers to these questions?
What was the date of the Indenture?
Who was King of England at that time?
How old was Elizabeth?
In which parish did she live?
Who was to be her employer?
How long was she to serve him?
What did Elizabeth's employer have to provide for her?
Find his signature and seal at the foot of the document.

Apprentices like Elizabeth Smith were herded together in lodging houses close to the mills and often did not get enough to eat.

In this picture a group of these ragged, hungry children are searching for food in a pig trough they have discovered. The boys in the doorway are pleading to be allowed to join in.

Below, two children are being sent out to work in the nearby cotton mill. You can see it through the window. They, too, look underfed and poorly dressed, and are obviously afraid of the woman. Can you find the bunch of keys hanging from her waist? Probably she locked up all the food in the house.

Notice the bare wooden floor and the simple furniture. The clock on the wall has a pendulum which swings backwards and forwards as the minutes tick by.

Through the open door you can see the scullery with its shelves of dishes and heavy, iron pans. Find the brick copper in which the water for washing clothes was heated by means of a fire underneath.

Here are children at work in a cotton factory. The man is probably the factory owner. How old do you think the children are? Are they as old as you? At first, children were not supposed to work in the factories until they were six or seven years old, but in fact, some were only three or four years of age.

The children worked from early morning until late at night and in some cases they were beaten to keep awake. Sometimes they were so tired they fell into the machines they were minding and were badly injured. The worse treated children were the orphan apprentices who had no father or mother to take care of them. They were sometimes so unhappy they ran away, though there was nowhere for them to go. They wandered around the streets of the large towns, looking for scraps of food to eat and a doorway in which to sleep.

ABSCONDED

From his Master's Service, at
WEST HARTLEPOOL,
ON THE 23rd OF APRIL LAST,
A YOUNG MAN, NAMED
WILLIAM SMURTHWAITE,

19 Years of Age, Stands 5 feet 8 inches High, Dark Complexion? Long Dark Hair, has Marks of a Cut across the inside of Three Fingers of his Left Hand.

His Father, WILLIAM SMURTHWAITE, Mason, resides at Houghton-le-Spring; his Brother, JAMES, at Walker Iron Works, near Newcastle; and it is supposed that he may be in one of those Neighbourhoods.

A REWARD OF

Will be paid by his Master, WILLIAM JAMESON, Cartwright, to any Person giving such Information as shall lead to his Apprehension.

Arthur Street, West Hartlepool, May 13, 1853.

From the Office of J. PROCTER, High Street, Hartlepool, and Victoria Terrace, West Hartlepool.

Older apprentices sometimes ran away, too, as you can see from this notice. William Smurthwaite had not returned to his family. Perhaps he thought he would be quickly discovered if he went home. Why do you think he had run away? Remember his employer had to provide him with certain necessities. What skilled trade was his master teaching him?

William had broken his part of the indenture, or contract, which had been signed many years previously, and at his age he would have become a skilled apprentice. So his master was prepared to offer a reward for information which would lead to his discovery. How long had William been missing before the reward was offered?

Sometimes, if differences between a master and his apprentice could not be settled, the indenture could be cancelled by magistrates.

This picture is taken from a novel which was written to show the way children were treated in factories. The ragged, barefoot boy is being greeted by a friend who is visiting the factory. Notice the clothes of the woman watching them, and the small boy crawling from under the machine. Compare them with the well-dressed visitors talking to the overseer, or foreman.

In 1832 Parliament agreed to investigate the conditions under which children were working in factories. Many people were afraid to give evidence in case they lost their jobs. Below are a few of the questions which were asked Samuel Coulson, a tailor at Stanningley near Leeds, and the answers he gave. He had three daughters working in the mills.

'Q   At what time . . . did those girls go to the mills?

A   For about six weeks they have gone at 3 o'clock in the morning, and ended at 10, or nearly half-past, at night.

Q   What intervals were allowed for rest or refreshment?

A   Breakfast a quarter of an hour, and dinner half an hour, and drinking a quarter of an hour.

Q   Did this excessive term of labour occasion much cruelty?

A   Yes. With being so very much fatigued the strap was frequently used.

Q   Have any of your children been strapped?

A   Yes, every one.

Q   Had your children any opportunity of sitting during those long days of labour?

A   No.'

Despite the results of the investigation, conditions in many factories continued to be bad and there were still many cases of cruelty reported.

At Shrewsbury: 'I was strapped and buckled to an iron pillar and flogged.' (A boy aged 10.)

At Leeds: 'If I was five minutes late, the overlooker would take a strap, and beat me till I was black and blue.' (A boy aged 7.)

At Wigan: 'For being late I had a weight of 20 pounds hung from a rope round my neck as a punishment.' (A girl aged 9.)

Gradually, new laws were passed to improve conditions. In 1833 a Factory Act prohibited the employment in cloth factories of children under nine years of age and reduced the working hours of older children.

Accidents in factories were frequent, for little care was taken to protect the workers. In the picture above find the boy with the brush who is underneath one of the spinning machines in a cotton mill. He had the dangerous job of sweeping up the waste materials while the machine was still working. Of one factory in Wolverhampton it was reported:

'The rooms are all crowded with dangerous machinery; so close that you can hardly pass . . . Not any of this machinery is boxed off, or guarded in any way . . . . You have but to once stumble on your passage from one place to another, or to be thinking of something else, and you are certain to be punished with the loss of a limb.'

The Factory Act of 1844 made it compulsory for all machinery to be fenced, and prohibited any woman or child from cleaning a moving machine.

A few factory owners tried to improve the working and living conditions of their employees. One of these was Robert Owen, a mill owner at New Lanark, near Glasgow.

This picture shows one of the villages Owen built around his factories so that his workers could live in healthy surroundings. Look for the trees and open spaces between the rows of houses and compare it with the second picture on page 42.

Robert Owen reduced the working hours of his employees to 12 hours a day and stopped employing poor apprentices and children under ten years of age in his factories. In 1816 he started a school which children from the age of three could attend until they were old enough to go into the factory. His workers could attend evening classes, too, and the picture below shows a dancing class being held for some of his young workers. Find the teacher on the right of the picture, the people sitting round the room watching, and the pictures on the wall which shows the room is used for other lessons during the day.

The document opposite is a page from a diary written by John Ward, a weaver who lived and worked at Clitheroe in Lancashire in the middle of the 19th century. Can you read what he has written? The American war he mentions caused a shortage of cotton which resulted in much unemployment in the Lancashire cotton mills.

John Ward had had a difficult winter. Why was he now unable to work and how much money had he to live on each week? Had it not been for the help he received he would have starved. In those days there was no social security payment for those who were sick or unable to work. Three shillings (15p) a week was barely enough to buy bread and a few other essential foods. Even after John Ward started work again he earned less than a shilling a day. What do you think he meant by 'firing'?

His diary was written in an old cash book and it was only by chance that it was discovered in 1947, almost a hundred years after it was written. A labourer picked it off a heap of rubbish which was being thrown into a furnace, realised it was of unusual interest, and reported his find. As its owner could not be traced, it was placed in the Lancashire Record Office.

Old diaries and papers are always interesting for they tell us something of the period in which they were written. Perhaps something which you write will be read with great interest in a hundred years time.

John Ward made the following entry in his diary a week later:

'I have had another weary week of bad work. I have just earned 7/3½ off three looms and there are plenty as bad off as me, and if any one complains to the Master of bad work he says, if you don't like it you can leave. He wants no one to stop that does not like it, and that is all the satisfaction we can get.'

1864

April 10 It is nearly two years
since I wrote anything in the
way of a diary I now take up my
pen to resume the task it has
been a very poor times for me all
the time, owing to the American
war which seems as far of being
settled as ever the Mill I work in
was stopped all last winter during
which time I had three shillings
per week allowed by the relief
committee which barely kept me
alive when we started work again
it was with Surat cotton and a
great number of weavers can only
mind two looms we can earn very
little I have not earned a shilling
a day this last month and there
are many like me my clothes
and bedding is wearing out very fast
and I have no means of getting
any more as what wages I get does
hardly keep me after paying rent
rates and firing I am living by my

Some things to do

This map shows how the populations of certain towns increased during the Industrial Revolution. Make a block graph in your scrapbook from the information the map contains. What do you notice about the position of the towns with the greatest increases? Which three places on the map had the smallest increases in population? Why do you think this was so?

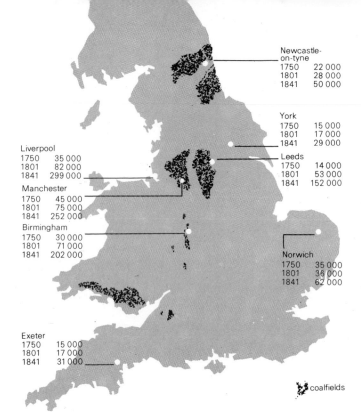

Newcastle-on-tyne
1750	22 000
1801	28 000
1841	50 000

York
1750	15 000
1801	17 000
1841	29 000

Liverpool
1750	35 000
1801	82 000
1841	299 000

Leeds
1750	14 000
1801	53 000
1841	152 000

Manchester
1750	45 000
1801	75 000
1841	252 000

Birmingham
1750	30 000
1801	71 000
1841	202 000

Norwich
1750	35 000
1801	36 000
1841	62 000

Exeter
1750	15 000
1801	17 000
1841	31 000

coalfields

Find out more about the following from encyclopedias and other reference books in your library and write notes about them in your scrapbook. Remember to illustrate your work.

John Kay	Samuel Crompton	James Watt
James Hargreaves	Thomas Newcomen	Abraham Darby
Richard Arkwright		

If you have a model steam engine at home ask your teacher if you may bring it to school and demonstrate how it can be used. Remember to warn your friends that steam is very hot and can scald. Point out the safety-valve and show them how it works. Why is it necessary to have a safety-valve?

Here are some museums which contain steam engines:
The Science Museum, South Kensington, London,
The Museum of Science and Industry, Birmingham,
The Sunderland Museum,
The National Museum of Wales, Cardiff, (models)
The Royal Scottish Museum, Edinburgh,
The Hunterian Museum, Glasgow University.
Visit your local museum and make drawings of any steam engines you find there. Label your drawings and write notes about them.

The Wedgwood potteries

Look at these lovely pieces of pottery. They were made in the Wedgwood potteries in Staffordshire during the 18th century.

Josiah Wedgwood came from a family of potters and lived at the time when factories were beginning to replace small groups of craftsmen. In 1759 he opened his own works at Burslem, where the nearby coalfields provided fuel for the kilns. He began to use new methods in the production of pottery and first-class artists to create new designs. Previously, pottery used by ordinary people had been coarse and dark in colour, and only the rich could afford porcelain. Wedgwood's new techniques produced a fine, cream-coloured earthenware which he called *Queen's Ware* and which was suitable for general use. Look carefully at the items of *Queen's Ware* at the bottom of the page.

He realised the need for improved transport and supported the building of better roads and canals, so that goods and materials could be carried more easily to and from the many factories which were opening in the Midlands.

In 1769 he built another factory near Hanley, alongside the partly built Grand Trunk Canal. Within a few years, Wedgwood was able to bring china clay from Cornwall by sea and along the new canal, and send his finished pottery to the ports of Liverpool and Hull.

Wedgwood pottery became so popular that showrooms were opened in London, Bath and Dublin, and fashionable people came from all over Europe to visit them and place their orders. By 1800 Wedgwood pottery was being sold throughout the world.

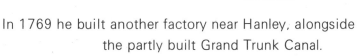

Clerk of the Manufactory ~~of Goods~~

To be at the works the first in the morning, & settle the people to their business as they come in, — to encourage those who come regularly to their time, letting them know that their regularity is properly noticed, & distinguishing them by repeated marks of approbation, from the less orderly part of the work people, by presents or other marks suitable to their age &c.

Those who come later than the hour appointed should be noticed, and if after repeated marks of disapprobation they do not come in due time, an account of the time they are deficient in should be taken, and so much of their wages stopt as the time comes to if they work by wages, and if they work by the piece they should after frequent notice be sent back to breakfast-time.

When the people are settled to their business, the clerk should then visit as many of the particular workshops as he can before breakfast time, and examine into the following particulars.

That the shop is clean & neat, and no heaps of dirt & rubbish swept into corners.

That the moulds, tools & furniture are in their proper places, & kept up good in their kinds, the moulds not over-worn &c.)

Read the above instructions given by Wedgwood to one of the foremen in his factory. You may think that Wedgwood was a very strict employer. He was, but he was also a very fair one and did all he could to keep his workers contented. He built houses for them, looked after them during illness and was kind to children in his employment. He was an efficient businessman and was constantly searching for new techniques to improve the manufacture of pottery and to develop other industries in the area.

Wedgwood and his friends, among whom were James Watt and Matthew Boulton, met once each month in each other's homes to carry out experiments with electricity and chemistry and to talk about them. They met at full moon so they could see their ways home. Can you understand why they called themselves the Lunar Society?

Housing

There was still plenty of work for some skilled craftsmen. The factory owners wanted houses built quickly and cheaply for the workers, as well as fine houses for themselves.

Above is a picture of a bricklayer at work. Notice the way he is dressed. What has he got in his right hand? What is his labourer doing? Find the hod leaning against the wall. This was used by the labourer for carrying bricks and mortar to the bricklayer. The outside wall of the house he is building is not very thick. Can you think why this is so?

The picture above shows plumbers at work. One man is melting solder in a ladle over a fire. Solder was used for joining lead pipes together. Another man has a coil of lead piping over his shoulder as he waits his turn at the fire. A young apprentice is carrying lead on his shoulder as he climbs up the ladder. Lead was used for gutters, cisterns and water piping, and also as a roof covering.

Some of the roofs were made of slates. The picture above shows men working in a slate quarry. They are splitting and shaping slates from large blocks which have been blasted from the face by gunpowder. Then the heavy slates are lifted into the cart and carried away.

Large quantities of lime were needed, too, for the bricklayers' mortar. Below is a picture of a lime kiln. Find the picks and hammer with which the men broke up large pieces of limestone. The two men have just emptied a basket of broken limestone into the kiln. At first, the kilns were heated by a wood fire, but later, coal was used. After heating, the powdered lime was carried away on horseback in barrels slung each side of the horse's saddle.

Look at this finely sculptured 18th century tombstone which stands in the churchyard at St. John-sub-Castro at Lewes in Sussex. What was the craftsman buried there? Make a list of the tools shown on the tombstone.

In this picture you can see a carpenter at work at his bench. Notice the way he is dressed. Can you see what he is doing? The curled shavings should give you a clue. There was plenty of work for him to do. Can you think of some of the things he would have to make for the many new houses?

Visit a local building site and find out as much as you can about the work of a carpenter nowadays. Has it changed at all? How many of the tools shown in the picture are still used?

During the 19th century unplanned towns quickly sprang up and spread like an ugly rash across the countryside.

Look at the picture above of the industrial town of Halifax. Notice the many tall factory chimneys and the cloud of grimy smoke which hung over the town. In many towns there were rows of small terraced houses near the factories. Many of them had no water or sanitation and people had to draw their water from a pump at the end of the row.

In the houses shown above, the walls of the back rooms also formed the backs of another row of houses built on behind them. These back-to-back houses had no backyards and only one door leading to the street or courtyard. Why was this dangerous? Compare the street lamp with those in your street. The one in the picture was lit by gas. How are yours lit today?

The rows of houses were overcrowded and some of them contained several families. In this picture a family with four children is sharing one basement room, which is all they have for their home. They have no curtains and a piece of sacking is hanging across the window to keep out the draughts. There is only one bed for the whole family, and a few pieces of bare furniture.

How is the room lit and what do the family use for heating and cooking? The man and his son look tired after their hard day's work. Probably the woman and her elder daughter have been working a long day at the factory as well, but there is no rest for them. What are they doing? Most likely the little girl playing with her baby sister has been looking after her all day.

It is not surprising that many children living under these conditions died of various diseases. Many adults, too, died when still quite young. Below is a table comparing the average age of death of factory labourers with gentlemen in four industrial towns about 1840.

	Labourers	Gentlemen
Bolton	18 years	34 years
Leeds	19 years	44 years
Liverpool	15 years	35 years
Manchester	17 years	38 years

Below is the budget of one Manchester housewife for a week in May, 1844. Her husband was a spinner in a cotton factory and at the end of a 72 hour week, brought home 20s 6d (£1·02½), while her daughter, aged 14, earned 4s 6d (22½p) a week at the same factory.

Butter, 1½lb at 10d	1s	3d
Tea, 1½ oz		4½d
Bread (cost of flour and salt)	4s	6d
Half a peck of oatmeal		6½d
Bacon, 1½lb		9d
Potatoes, two score a week	1s	4d
Milk, a quart a day, 3d a quart	1s	9d
Meat, 1lb on Sunday		7d
Sugar, 1½lb a week at 6d		9d
Pepper, salt, mustard and extras		3d
Soap and candles	1s	0d
Coal	1s	6d
Rent of house per week	3s	6d
	———	
	18s	1d (90½p)
Clothes, sickness, etc.	6s	11d
	———	
Weekly income	25s	0d (£1·25)
	———	

This family, like other poor families, made their own bread, and they would taste no other vegetable than potatoes.

Each week 1d was paid to a funeral society for each child.

Two of the children went to school for 3d a week. For this they were taught reading, but not writing.

This was an average budget of a spinner's family, but, as nowadays, wages varied according to the type of work. An unskilled labourer earned much less and his wages were fixed according to the cost of bread. The working man had no money for luxuries, whereas the employers became wealthier as trade increased.

Below you can see the kind of meals such a family would eat.

Breakfast: Porridge, bread and milk
           On Sunday — cup of tea, bread and butter
Dinner: Potatoes, bacon and bread
           On Sunday — a little meat; no butter, egg, or pudding
Tea: Tea, bread and butter
Supper: Oatmeal porridge and milk; occasionally potatoes and milk
           On Sunday — sometimes a little bread and cheese

Conditions in the mines

In places where no other power was available, horses were still used to drive wheels. At some pit-heads they turned the winding gear which brought up coal from the bottom of the shaft and carried miners to and from their work at the coal face.

Look carefully at the picture above of a horse-gin in action at a pit-head. Notice how the horses are harnessed at each end of the long bar. Can you see how the horses worked the winding gear by walking round and round in a circle? The man with the whip made sure the horses kept moving. Find the group of men who are unloading coal which has been brought to the surface. What is the man with the pick-axe doing?

As you can see from the picture on the left, it was back-breaking work at the pit-head, for there was no machinery to help the labourers. Here they are weighing a tub of coal.

This chart shows how the production of coal increased between 1770 and 1850.

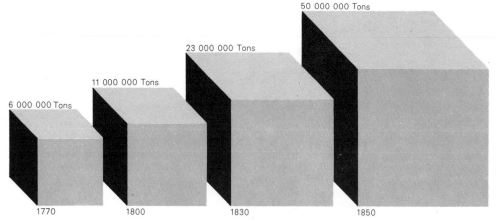

More and more coal was needed for the steam engines to drive the factory machinery, and many new factories were built in the mining areas. Coal was needed, too, for the railways being built during the early 19th century. But conditions in the mines were even worse than those in the factories. Men, women and children all worked long hours underground and their work was hard and dangerous.

The picture above shows a woman called Betty Harris working underground in a colliery near Bolton in Lancashire. She is hauling a heavy tub of coal up a slope by pulling herself along with the aid of a rope. She has a boy to help her with her task. Find the chain from the tub which is fastened to a belt around her waist. The tunnel is not tall enough for her to stand upright. This is what Betty Harris said about her work:

'The pit is very wet where I work, and the water comes over our clog-tops always, and I have seen it up to my thighs; it rains in at the roof terribly. My clothes are wet through almost all day long. I am very tired when I get home at night. I fall asleep sometimes before I get washed.'

The miners' lives depended on a flow of fresh air, controlled by means of trap-doors. Young children called trappers, some only five years old, sat waiting at the trap-doors for the coal trucks, or 'corves', as they were called.

Here is a picture of a trapper sitting on the cold, damp floor. He looks as if he has fallen asleep, which is not surprising, for often he sat there for 12 hours a day. Behind him you can see a hole scooped out of the wall, in which he crouched when the corve approached. He pulled the door open with a piece of string to let the corve pass through, and then let it close again.

It was not difficult work, but very important, and the children were cold, frightened and lonely as they sat there, usually in darkness. Girls as well as boys were trappers, and one eight-year old girl fell asleep in a pit in South Wales. When she awoke, she found that rats had eaten her bread and cheese.

Another little girl of eight said:
'I'm a trapper in the pit. It does not tire me, but I have to trap without a light and I'm scared. Sometimes I sing when I'm frightened.'

Although steam engines were used in many pits to raise the coal to the surface, women and girls were still employed as coal bearers in some places, particularly in Scotland. They were cheaper than machinery, especially where the output of coal was small.

Here are two girls carrying baskets of coal up a ladder. This was dangerous and exhausting work. Sometimes a loaded basket was so heavy that it took more than one man to lift it on to the girl's back, for often over 2 cwts at a time were carried in this way.

Look closely at the picture and you can see lumps of coal resting on each girl's neck and shoulders as she stooped to carry her heavy burden. The basket was strapped over the girl's head to keep it in place. Sometimes the straps broke and the load of coal fell on the girl below her, knocking her off the ladder, which was about 6 metres high. You can see what has happened to one large lump of coal carried by the first girl. What do you think has caused it to fall?

The girls had to climb several of these steep ladders and walk along underground passages which were only about 1 metre high, before they reached the place where they could get rid of their load. Then it was time to walk the long way back to collect the next one. Sometimes they worked like this for ten hours without a rest, and for this they were paid about eightpence (3p) a day. Women and girls of all ages from six to sixty carried coal on their backs in this way.

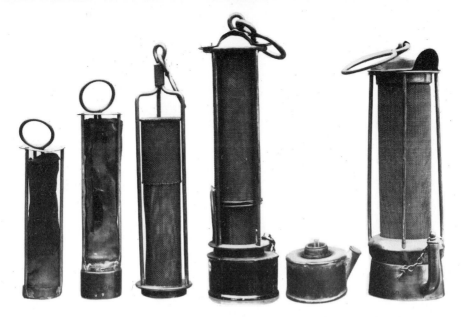

At first the only lighting in the mines was by candles, and many deaths were caused by explosions of the coal-gas known as 'firedamp'.

In August, 1815, a scientist named Sir Humphry Davy was asked to invent a lamp which would be safe to use, and which would give the miner warning of danger. Some of his first attempts are shown above.

Davy discovered that if the flame of a lamp was surrounded by a cylinder of fine wire gauze and closed at the top, it was safe to use even when there was inflammable gas nearby.

In January, 1816, the two oil lamps shown on the right were tried out at Hebburn Colliery in County Durham. Can you see where they were filled with oil? The flame would not pass through the wire gauze except in a strong draught. If coal-gas passed through the gauze into the lamp, the flame went out and the gas inside the lamp was set alight. This warned the miner of the danger and gave him a chance to escape.

These first successful lamps can now be seen in the Science Museum in London, but improved models were soon used in many coal mines.

By the middle of the 19th century conditions in the mines were gradually improving. Women and children were no longer allowed to work underground and ponies were being used to pull the heavy loads along.

Look at this busy scene at the bottom of the shaft of a coal mine in Staffordshire and find:
— the 'hewers' hacking out the lumps of coal with their pickaxes. Notice how pillars of coal have been left in place to hold up the roof.
— the man carefully loading the trolley with coal
— a loaded trolley on its way up the shaft on the right of the picture. How do you think it was being pulled up? Notice how metal hoops prevent the coal from falling off the trolley.
— the ponies pulling the trolleys along the rails. Why do you think they wear 'blinkers'?
— the naked candles on the left of the picture, despite the invention of Davy's lamp some years earlier.

In general, the miners were better paid than other workers and therefore were better clothed and housed. The miners' cottages in the picture above were at Long Benton in Northumberland and were much better than those in other parts of the country, though we should not find them very convenient today. Look closely at the picture and find:

— the row of outhouses (sheds, lavatories and washhouses) facing the cottages
— the pit-head in the background
— the bridge across the little stream and the large barrel for collecting rain water from the roof
— the tall, smoking chimneys. The miners had plenty of coal to keep their houses warm.

Sometimes part of the miners' wages was paid in tickets which could be exchanged for goods only at 'truck shops' or 'tommy shops'. Many of these were run by the foremen or even the employers themselves, and prices in these shops were much higher than anywhere else. Even when a miner was paid his full wages in cash he was expected to buy goods at the tommy shop if he wanted to keep his job. Wages were not always paid regularly and sometimes miners had to wait several weeks for them. Their wives were forced to obtain goods on credit from the high-priced tommy shops and pay for them later when their husbands received their wages.

Early trade unions

During the 18th century groups of workers began to fight for better wages and working conditions. Can you think why most employers were against this?

In 1799 Parliament passed the Combination Acts, which allowed employers to fix whatever rates of wages they liked and made it illegal for workers to hold public meetings. Ten years earlier, workers in France had revolted against the rich ruling classes and Parliament were afraid of riots in this country. The workers continued to meet and many were arrested and sent to prison.

This picture shows a large crowd assembled at St. Peter's Fields in Manchester to listen to a speaker called Henry Hunt. The magistrates called in mounted soldiers to arrest the speaker and clear the crowd. The horses became nervous of the huge crowd and the soldiers began to panic and strike out with their drawn swords. Notice:
— those who have been struck down or trampled underfoot
— the speakers on the platform and the banners being displayed
— the soldier attacking one of the banner bearers.
About 15 people were killed and hundreds were injured. This incident later became known as the Peterloo Massacre.

In 1825 another Act of Parliament allowed workers to meet together, but only to discuss wages and hours of work. By 1830 workers were trying to form societies or unions for their fellow workers in their particular trades. Members paid weekly fees which enabled some societies to provide insurance against absence from work through injury or sickness.

This is a membership card of one of these early trade unions. In this case, it had only women members. What was their work? Notice the motto and picture at the top of the card.

Robert Owen was one employer who supported these early attempts to form trade unions. (Can you remember what else he did to help his employees? If not, look back at page 33.) Owen thought that if these societies of skilled workers joined together to form one large union they could 'strike' (stop working). Then employers would be forced to consider their complaints.

So in 1833 the Grand National Consolidated Trade Union was formed, but it was not a success. Even though trade unions were now legal, some employers dismissed workers who were union members. Others called out soldiers to deal with strikes and riots. There were many angry scenes and rioters were arrested. Some were imprisoned, some even executed and others were transported to convict settlements in Australia.

But by the end of the 19th century there were many unions of both skilled and unskilled workers, and employers and workers began to discuss with each other working conditions and wages.

Communications

It was essential that continuous supplies of coal should reach the factories. Without coal the factory machines would come to a halt. Raw materials, too, had to be brought to the factories and the manufactured goods taken away to be sold.

Pack-horses like these were used in the 18th century to carry heavy loads along rough tracks and roads, especially in hilly districts. Even coal was carried in this way.

Look at this heavy wagon used for carrying goods over long distances during the 18th and 19th centuries. Notice the broad wheels, compulsory by law, which helped to roll out the many bumps and ruts in the rough roads. The transport of heavy and bulky goods by road was slow and expensive. The pack-horses could travel only at a walking pace and the big, lumbering wagons could travel only about ten miles (sixteen km) a day.

Some goods were sent along the rivers, but many of these were not deep enough to take the boats and others dried up during the summer months. Other goods were sent by sea round the coast of Britain.

London's coal came by sea from Newcastle upon Tyne and Sunderland. This picture shows a sailing ship leaving Sunderland harbour with a cargo for London. The men turning the capstan are moving the ship so that the wind can fill the sails.

Coalmine owners soon began to look for cheaper and quicker methods of carrying coal.

1761   An engineer called James Brindley completed a canal which enabled large quantities of coal to be carried from the Duke of Bridgewater's mine at Worsley to Manchester, about seven miles away (11 km), more quickly and at much less cost than before. The price of coal in Manchester fell by half. How would this affect the factories?

1762–   A network of canals was built across the country, joining rivers and linking
1820   them with the factory towns and ports. Because of its position, Birmingham became an important industrial centre.

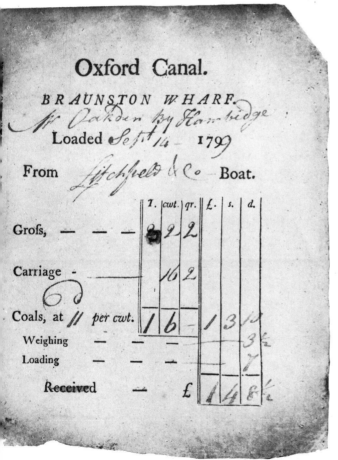

Horse-drawn barges hauled heavy loads of coal, iron and other raw materials to the factories at a quarter of the cost of previous transport.

Look at this receipt for a load of coal carried on the Oxford Canal. The charges were based on the weight of the load and the distance it was to be carried.

Notice the additional charges for weighing the loaded barge and for loading the coal.

You have read how factories and houses sprang up near the coalfields and iron foundries. Businessmen now built new factories alongside the canals. Coal and raw materials could be brought cheaply and easily to them and the goods they made could be sent along the canals to other parts of the country and to the docks. Can you remember the potter who extended his business in this way? If not, look back at page 37.

You will remember from page 20 that in 1781 James Watt's steam engines had been able to turn wheels and drive machinery. Men now began to think how steam power could be used to pull wagons along a track, for this would enable coal, raw materials and manufactured goods to be moved more quickly.

1804   An engineer called Richard Trevithick built the first steam locomotive to run on wheels. It hauled a train of wagons along a railway in Merthyr Tydfil in Wales.

1812   John Blenkinsop built a 'steam carriage' to pull the coal wagons along a track from Middleton to Leeds.

1825   The first public steam railway to carry both goods and passengers was opened and ran between Stockton and Darlington. An engineer called George Stephenson had been asked by the pit owners in South Durham to build the railway so that coal could be transported to Stockton more cheaply than by road.

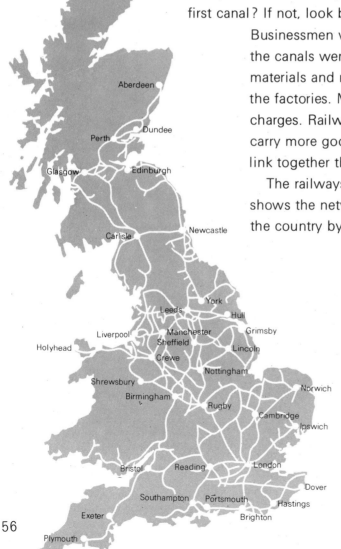

Can you remember why Brindley was asked to build his first canal? If not, look back at page 55.

Businessmen were beginning to realise that the canals were too slow for moving raw materials and manufactured goods to and from the factories. Many canals had increased their charges. Railways were much faster and could carry more goods, so routes were planned to link together the big cities.

The railways spread rapidly and this map shows the network of railways that covered the country by the middle of the 19th century.

By then, ordinary people were able to travel easily about the country and goods could be sent more quickly than ever before. The canals carried less and less traffic and many were sold to the railway companies, while others closed down.

But the railways consumed vast quantities of coal themselves, so even greater supplies were needed from the coal mines.

This diagram shows how coal and iron were taken by the railways to the factories and ports, and also how the railways and steamships were themselves dependent upon large supplies of coal.

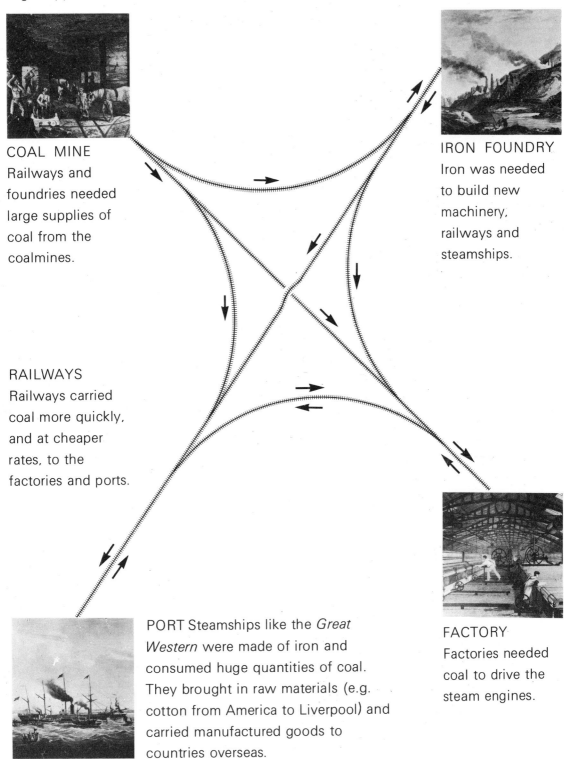

COAL MINE
Railways and foundries needed large supplies of coal from the coalmines.

IRON FOUNDRY
Iron was needed to build new machinery, railways and steamships.

RAILWAYS
Railways carried coal more quickly, and at cheaper rates, to the factories and ports.

PORT Steamships like the *Great Western* were made of iron and consumed huge quantities of coal. They brought in raw materials (e.g. cotton from America to Liverpool) and carried manufactured goods to countries overseas.

FACTORY
Factories needed coal to drive the steam engines.

Here is a picture of the port of Barrow-in-Furness during the 19th century. Look at the many factories and warehouses on the far side of the river and the clouds of thick, black smoke pouring from the tall factory chimneys.

In the front of the picture a woman and her children have left their dog to guard their belongings while they talk to the man sitting on a log.

How many kinds of sailing ships can you find? Can you see the two steamships in the picture? Notice their funnels and the large paddle wheels which drove the ships along although they still had sails as well.

In 1838 a large paddle steamer called the *Great Western* sailed across the Atlantic and soon afterwards a regular steamship line was carrying goods and passengers between Bristol and New York.

So the Factory Age led, not only to great changes in industry and to many people's way of life, but also to a revolution in transport. The new and better roads and the canals and railways reduced transport costs and increased trade, while the steamships brought the far countries of the world within reach of Britain.

Although it was a time of new inventions and industries, skilled craftsmen were being replaced by machines. In some parts of the country green fields were replaced by sprawling, ugly, overcrowded towns and the fresh air of the countryside gave way to a dirty, smoke-laden atmosphere from factory chimneys.

The picture below shows a view of Stockport in Cheshire in 1848. Look at the unplanned muddle of factories, houses and transport. Dirty, industrial waste water is pouring into the river from the factories, and smoke from house and factory chimneys pollutes the atmosphere. How many different ways are goods being carried along the road on the left of the picture? What can you see on the tall viaduct which crosses the river and road?

Industrial archaeology

Many people are interested in discovering what still remains of the Factory Age. The study of early factories, machinery, railways, canals and other industrial features is known as industrial archaeology. If you search in your district and also in your local museum and library, you may be able to discover for yourself further information about the Factory Age.

Are there rows of terraced houses like these in your nearest town? If so, they were probably built to house workers in nearby factories or mines. Some buildings have the year in which they were built shown on their brickwork. Try to find any in your district which were built during the 18th or 19th century.

Go and look at the factories in your town. Even though they appear to be modern there may be an older block at the rear of the building. If so, try and find out what it was first used for.

Sometimes the buildings are no longer being used for the purpose for which they were built. This former cotton mill at Langworth, in Nottinghamshire, for example, is now being used as a barn.

If you live in a mining area you may be able to find evidence of former mines which have been closed down. There may be remains of miners' cottages, or brick sheds where the coal was washed and sorted. The heaps of waste coal slag may have been grassed over, but sometimes the tall, narrow building which housed the steam winding engine may still be standing. The one above is at Beamish in County Durham. Notice the winding gear which pulled up the heavy loads from the mine below.

There may be a museum of early mining tools and equipment for you to explore. If so, look for safety lamps, helmets and boots.

Look at the warning bell above which was made about 1770. It was used to call a rescue team when a disaster occurred in the mines at Leadhills in Scotland.

This picture shows all that remains of Wedgwood's canal-side factory at Etruria. There may be remains of old factories in your nearby towns. If so, try to find out what was made in them.

Even if there are no mines or factories in your area you may be able to discover interesting stories about other things. Look for old bridges, manhole covers, lamp posts, letter boxes, pumps, railings and milestones. Find out all you can about them.

Main achievements during the Factory Age

1712   Steam engine built by Thomas Newcomen

1733   Flying shuttle invented by John Kay

1750—

1758   First Act of Parliament to authorise a railway

1759   Josiah Wedgwood opened his pottery at Burslem

1761   James Brindley completed the Bridgewater Canal

1767   Spinning jenny invented by James Hargreaves

1769   Richard Arkwright invented spinning machine

         Josiah Wedgwood opened factory at Hanley

1779   Spinning mule invented by Samuel Crompton

         Iron bridge built over the River Severn

1781   James Watt built a rotative steam engine

1785   Power loom invented by Edmund Cartwright

1800—

1804   Richard Trevithick built first steam engine to run on wheels

1812   Steam locomotive built by John Blenkinsop used at Middleton Colliery

1816   Safety lamps invented by Humphry Davy used at Hebburn Colliery

         School started by Robert Owen for children from age of three

1819   Act of Parliament stopped children under nine working in cotton mills

1825   Stockton to Darlington Railway opened

1830   Liverpool and Manchester Railway opened

1838   Paddle steamer *Great Western* sailed across the Atlantic

1844   Factory Act compelled all machinery to be fenced and prohibited women and children from cleaning a moving machine

1845   The first iron ship crossed the Atlantic

1850—

Things to do

Now that you have finished reading this book you will be able to add some more pictures and notes to your scrapbook. Here are some things for you to do:

Find out which new industries grew up in your district during the 18th and 19th centuries. If any houses were built for the workers, draw a map to show their position and make sketches of them.

Write a story about Ellen Turner, an orphan girl of ten, who was sent to work in a Lancashire cotton mill as an apprentice. Describe her work at the mill, how she was treated by her employer, and her poor living conditions.

Paint large pictures of:
Children at work in a factory.
Women hauling tubs of coal at the bottom of a mine shaft.
An overcrowded basement in a factory town.
A canal scene.
An industrial port.

Here are the names of some famous people to look up in your encyclopedia:

Robert Owen	Thomas Telford	George Stephenson
Earl of Shaftesbury	James Brindley	Isambard Kingdom Brunel
John Macadam	Richard Trevithick	

Write notes about each of them and make drawings to illustrate the work they did.

Imagine you are an eight-year-old boy working down the mine as a trapper. Tell your friends what happened and how you felt during your first week's work.

Paint a wall frieze called 'The Factory Age' for your classroom. Include rows of terraced houses, factories with smoking chimneys, and a canal.

Index